For

# Elliott & Grace

*may they add to this list*

© *Stephen Griffiths* January 2014
© Cover Illustration by *Genny Haines*
Acknowledgement **Design Cuts**

# Context

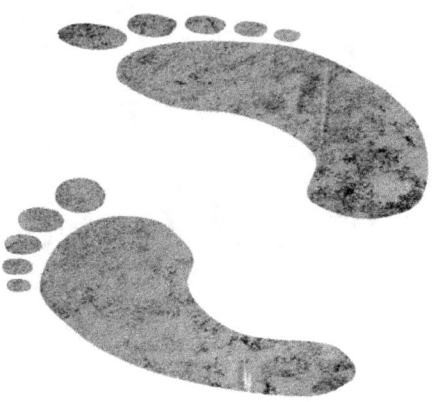

There is a universal belief that science enjoys an ever increasing rate of discovery, with more new, life-changing and ever-improving products being developed today than at any time in human history. It's one of those beliefs held so strongly that it's almost never questioned, let alone tested.

Yet only a few areas of activity have changed significantly in the past few hundred years and generally these improvements are available to only a small segment of the earth's teeming populations. When we look at the inventions that have significantly enhanced either the quality or quantity of a typical human life, there have been only marginal improvements since the invention of writing a few thousand years ago.

*The real life changing inventions and key discoveries were made thousands or tens of thousands of years before written records began.*

The fact is that most of the modern world's population, 7 billion souls, will invent nothing that contributes to our science and technology. This is despite the invention and application of lengthy and expensive education for our youth and a world wide computer web that bring almost everything to anyone's fingertips for close study.

Much of the experience of modern life is out of touch of the natural world.
We don't even think about the genius of our distant ancestors.

Try the following thought experiment:

Put a modern family on a fairly hospitable, but temperate, island.
Be generous and give them enough fresh food to last a full two years, unlimited clean water and a fairly abundant flora and fauna to exploit.

How long would they survive?

The surprising answer is that they are unlikely to see out the first winter.

Before explaining this claim, let's remind ourselves of what they *didn't* have:

They were given no clothes, no seed stocks, no tools, no ropes or nets, no chemicals, no pets and no medicines.

They started with plenty of food, but are unlikely to know how to preserve it from dampness, moulds, insects and rodents.

Modern farming uses especially created dry storage units for grain and chemical and physical preservatives for fruit and vegetables, including anti-microbial treatments, sprays and freeze-drying. An often overlooked type of preservative is the various wrappings and packages food is stored in.

Yet despite the billions spent on food storage and preservation, public health still allows a significant percentage of rodent hair and insects parts in food (Ohio University estimates that the average American eats nearly two pounds of insects unknowingly every year).

And still millions of tons of foodstuffs are lost annually in spoilt produce.

That quantity excludes all that wasted post purchase, because it's uneaten and left in the packet or discarded before the sell by date. This waste is estimated at nearly a third of all food bought. Could our isolated, modern family do better?
Would they even know how to make clothing from the natural resources all around them and how to stay warm and healthy?
Could they create and develop a sustainable environment for themselves?

And would they achieve all this *in time*?

They would need to know what wild plants were safe to eat, how to trap game and fish and how to preserve what was harvested for the long lean winter to come. They would also have to stockpile an adequate quantity of fuel for their fires before the weather became too extreme to venture out.

And, should they fall sick or be injured, they would need a basic knowledge of first aid and medicine. They would all need to stay warm when the thermometer plunged below zero.

The best solution to the isolated family's predicament is much the same as it was for our oldest, pre-stone age ancestors - skip out of the temperate zone and stay in the tropics.

Pre-humans could only survive where there was no great seasonal variations and food was so abundant that they didn't need to preserve anything for very long. They were tied closely to water-sources such as rivers and lakes and restricted to the tropics.

A million years ago mankind could live almost nowhere else on earth. Like chimpanzees and the great apes, our home ranges were limited to the ecological niches we had specifically evolved to exploit and survive in, while trying to avoid being on the menu for the larger predators or those that could hunt in packs.

*Our human form has just two intrinsic advantages over most rivals. Fit humans can run for longer than most potential prey and from a stationary start we have excellent acceleration, reaching our maximum speed in just three paces. Despite the former attribute, there are no known examples of primitive people hunting by running down their prey over long distances.*

*And we have numerous disadvantages. Easily cut and hurt, relatively slow, no sizeable killing teeth or talons. Below average hearing abilities, poor night vision, easily chilled or burnt... the list goes on.*

*And whatever activity modern humans evolved to perform, it wasn't climbing trees. Our arms are so weak that even weightlifters and bodybuilders are vastly inferior, strength-wise, to every other large primate.*

# *Chapters*

**Prologue**    **Context**
**Dominant apes**

| | |
|---|---|
| Great Invention Number 1. | **The Missile.** |
| Great Invention number 2. | **Weaving & Knots** |
| Great invention number 3. | **Clothing and shelters.** |
| Great invention number 4. | **Floats & Boats** |
| Great invention number 5. | **Fire** |
| Great invention number 6. | **Food Preservation.** |
| Great invention number 7. | **Trade** |
| Great invention number 8. | **Animal Domestication** |
| Great invention number 9. | **Agriculture** |
| Great invention number 10. | **Sanitation** |

**Man's route to World Conquest**
**Author's Note**

# Dominant apes

There was no specific point in time when our distant ancestors became modern humans in a strict, anatomical sense. If the first true Homo Sapiens emerged 300,000 years ago (the fossil record is scanty and open to interpretation, recognisably modern Homo Sapiens evolved between 300,000 and two million years ago, depending upon which criteria is used), those ancestors could still mate successfully with proto humans from earlier generations and so on backwards to the very beginnings of life.

So wherever we draw the line and whoever came first, they inherited or acquired a certain skill-set and even a culture from their pre-human parents.
And we can easily deduce some of the assets and skills those creatures we now class as pre-human, gave to those first 'true' homo sapiens. By analysing the skill-sets of the apes and other mammals, we can deduce some of what the first proto-humans already knew.

And they would have known a great deal.

They'd know:

Which plants and animals in their immediate environment are food sources, when and where they are available and some idea of how nutritious they are.
How to evade and avoid the most dangerous local predators.
How to communicate with their colleagues in the immediate family, troop or tribe.
How to construct simple shelters.

How to use some very primitive tools.

There would also have been a division of labour and skills, both by age and sex. The likelihood is that the classic males-hunt and protect and women-gather and nurture was already well established, with the older family and tribal members being more highly skilled, knowledgeable and specialised.

The full details of these skill sets and attributes can never be fully known. For example, modern chimpanzees use some plants as medicines (eating certain leaves or bark to alleviate stomach complaints or remove parasitic infections), but which plants were harvested and at what point in their growth cycle is now impossible to determine.

It's therefore highly likely that some of our ape and pre-ape cousins were better pharmacists than most modern, urban humans.

With regard to language, we are still discovering the huge range of both verbal and non-verbal communication skills primates have at their disposal.

They undoubtedly had clear signals to communicate danger and identify predators by type (for example, many primates have different calls for leopards and snakes, some also include specialist calls for high-flying predators like eagles), the demanding feed-me messages of the young and invitations to groom or play and, not least, to mate.

The vocabulary a million or two years ago would have already exceeded some fifty or so sounds, plus various signals such as drumming and leaf tearing to suggest aggression or 'do not disturb' type messages. Troops of monkeys and apes will work together to intimidate predators or hunt meat, coordinating with their fellows by gestures and grunts. Apes construct nests in trees and other primates (especially orang-utans) are known to use large leaves as umbrellas.

Most importantly, many primates use tools to enable access to local foodstuffs. For example, chimps and capuchins use stones (even carefully selecting a suitable hammer and anvil) to open nuts, stick-wands to fish for insects and chewed leaves as sponges to soak up water. Chimps may also use fallen branches and stones to throw at leopards and other predators, although, compared to humans, they are not very good at throwing.

Which brings us nicely onto the first truly amazing human invention….

Great Invention No. 1

# The Missile.

*Development started about two million years ago.*

**The discovery of effective missiles predates the first true humans, beginning with simple sticks and stones in pre-human times and mastered with the coming of the bow and arrow some twenty five to perhaps as much as seventy five thousand years ago.**

The single biggest difference between ourselves and other large animals is that humans learned to kill at a distance.

For a primate that lacks the speed, strength, claws and the killing teeth of its predators, the importance of this discovery simply *cannot* be overstated.

Before missiles any large animal would be likely to injure those humans who came too close, and in a time before antibiotics any scratch from a dirty claw or unclean tooth could be fatal. Even a pulled or torn muscle could seriously reduce man's ability to hunt or gather food. An injured man is changed from an asset to his troop into a liability, a non-productive creature who needs resources like food, water and protection and slows down everyone else.

So missiles made hunting an activity that would be far less dangerous, less stressful and put far more meat on man's menu. And in a spin off from just being an aid to hunting, the development of ranged weapons is also one of the most socially and politically important discoveries.

Suddenly the tribal leader need not be physically strongest, he could be the most intelligent or the most social (the best at net-working and building relationships).

Missiles give mankind two huge advantages in their hunt for food:
I) The physical risks are massively reduced, and
II) the size and variety of prey animals increases.

In addition to the above, there are two really important indirect consequences:

Firstly, hunting with missiles offers increased opportunities for group activity, with volleys of missiles used to bring down large game. This will no doubt have necessitated some linguistic expansion.

Secondly, this access to missiles reduces the risk of predation. Large animals were either eaten, driven off or learned that we were no longer on their menu. In short, missiles made men the local apex predator. We would now dominate our environment, even if this was still fairly restricted geographically.

So how did this happen? How did relatively slow and weak homo-sapiens develop the tools to dominate their environment while other hominids did not?

Humans, males especially, have a shoulder joint design ideal for throwing.
Exactly why this evolved is unclear, but we do know that our shoulders are superior (for throwing) to those of *all* other primates. We also have the rare opposable thumb and finger combination that facilitates grip and release for hand-held weapons.

Given that proto-humans were highly dependent upon freshwater and coastal resources for food, perhaps our shoulders (and lack of body hair) evolved in response to spending a considerable amount of time in the water. We can swim much better than any other ape and our dependence upon water-based skills lead us to evolve, over many generations, a unique and versatile shoulder joint. The primate with a body hair pattern most similar to ours is probably the baboon, with only short hair from the waist down and, interestingly, many also have a fondness for sea foods like clams and shark eggs. Harvesting and opening shells may have been a crucial factor in the development of our thumbs.

Once we had learnt to swim and developed this unusual shoulder joint, it also gave us the capacity to throw missiles fairly accurately.

*Perhaps the first use of sharpened sticks was as harpoons for shallow water fishing.*

We were not apes that came down from the trees, we were apes that moved to the beach. My personal belief is that we were a minor hominid species driven from the tropical ape-heartlands by stronger competitors to eke out a marginal existence along the shores of seas, lakes and rivers.

What is particularly telling about the human throwing skill is that there is a significant difference between males and females. Young boys are generally two or three years ahead of the girls in throwing range and accuracy by the tender age of six. There is no such divide in swimming skill, which strongly supports the contention that our aquatic skills evolved before males were preferentially selected for throwing aptitude.

The fact that a sex-related, innate difference between male and female throwing skill exists almost guarantees it was, from an evolutionary perspective, a competitive advantage of vital importance.

The chain of discovery for missiles probably went something like this:

Proto-humans would hurl anything handy (sticks or stones) at something they wanted to drive off. They quickly found that curved or pointy sticks and certain stone shapes and sizes worked better. The first manufactured missile was probably a versatile, hand-hurled throwing stick similar to the Australian aboriginals boomerang or a stone age man's javelin. It would be used in defense and for hunting.

There have been numerous instances of modern scientists studying isolated tribes in the twentieth and twenty first century that didn't 'progress' much out of the stone age. We therefore had ample opportunity to study the technology of primitive men.

The 'stone age' label is a complete misnomer for a time that should have been more correctly called the 'organic age', after the materials that formed most of our early products.

These products were made from fibrous plants, hides, bone and wood far more often than stone, but such material doesn't preserve well. Indeed, peoples that live in jungles and dense forests leave very little material evidence of their existence. Some ancient African kingdoms have only been recently discovered because almost all evidence for their dwellings and buildings ends up being digested by termites.

Humanity moved along the road to world conquest by developing an aptitude for swimming, then throwing, and as a spin-off developed the first human science.
That branch of physics we know as aerodynamics.

Stick shape, size, weight and edge would be tested to produce the optimal missile for either accuracy (when hunting small game) or killing power when after something larger.

Throwing sticks and stones would be supplemented by something recognisable as a javelin or short spear. Such spears are wonderfully versatile weapons effective in close combat and over a short range, particularly against larger adversaries.

When we see vast herds of game on the plains of East Africa it's always surprising how close prey animals allow predators to approach, without resorting to the 'fight or flight' response that would take them out of harm's way.

So, what to modern humans is a very short ranged missile weapon (such as a javelin that can be cast around thirty metres or an arrow that can penetrate thick hides at up to a hundred), may still have had the scope to massively change our accessibility to fresh meat.

If an antelope will not run from a lion until it's within twenty paces, having a spear that can be thrown thirty will make for a relatively easy and very successful hunt. When such weapons were new large animals would have reacted like the now extinct Dodo, blissfully unaware of the danger from human predation.

Constant use would show that certain woods made for better spears and fire-hardening the points made them stronger, sharper and last longer.

The throwing spear was used on both land and sea to catch large animals and fish.
Initially just sharpened wood, extra penetration was achieved by fitting flint, obsidian, horn, bone and antler points.

The most widely known adaptation for hunting at sea was the harpoon (basically a spear with a barbed head and, later, an attached rope), which is thought to have been in use at least 80,000 years ago (in Zaire).

Yet there are significant limitations in the use of a javelin as a hunting tool. The size and weight of the spear restricts the hunter to carrying no more than a handful at a time. The user lacks concealment and must break cover when throwing the missile (as he has to stand and perhaps even run to maximise its range), and this range is limited by the length and power of the user's arm and overall upper body strength.

The flight speed of the thrown missile is also sufficiently slow that agile animals can dodge an accurate cast beyond twenty paces, or a skilled human opponent could catch the missile and even send it back.

Before the development of string or rope, the javelin's range was improved by a simple spear-thrower, a wooden stick that notched against the back of the spear and was used as an extension of the thrower's arm to give the cast greater leverage.

Any chimp can throw a stick. Only a man would invent a stick to throw his stick even further…

It doubled the effective range, although this extra distance came at the cost of a slower rate of fire. Providing the target was hit first time this wasn't important, while being further away from the target made the caster even safer. The throwing spear was thus a fantastic tool for hunting large game, but less effective at trapping the much more numerous small prey.

A throwing-spear armed hunter can expect to hit an antelope at thirty or more meters, but a small bird at no more than ten.

The development of the throwing spear and its further enhancement with the spear-thrower ensured the permanent place of human beings as the dominant predator throughout all our environment, by significantly enhanced hunting efficiency. A group of humans armed with throwing spears could take on any of the massive animals that prowled the pre stone age world. Mammoth, giant ground sloth, wild horses and even lions could be killed or put to flight.

At the same time that sticks were being developed into sophisticated ranged weapons something similar was happening with stones. The frustrating range limit for hand hurled stones and spears left a gap for a much more powerful weapon. This niche was filled by the sling, a simple tool to hurl stones up to and even over 400 metres.

A sling can be made from anything that will hold a stone and be strong and flexible enough to be spun around. A strip of hide or bark or a short length of cloth for example. And the discovery of the sling may well have been down to a chance event, rather than a particularly brilliant thought process.

Imagine a hunter collecting pebbles from a stream bed to use in herding game (such as wild horses) or for throwing at birds. He quickly realises he is more productive gathering the missiles in a bag, perhaps nothing more sophisticated than the skin of a small animal. He goes to the same stretch of riverbed until the supply of the better sized stones is becoming exhausted. Soon he is walking around a little frustrated, swinging the bag with just one or two stones in it. One slips out and suddenly there's the eureka moment, it goes a surprisingly long distance. Our hero now tries to replicate the action, and the sling is born.

It takes practise to completely master the sling, far more than is required for a javelin, but the sling's advantages over throwing spears can make this well worthwhile.
The sling increases the range of hurled stones roughly fourfold, adds almost unlimited ammunition and can be used to drive or herd prey by using the noise emitted when shot hits solid stone or wood. The high velocity and small size of slingshot also means it's less likely that even an agile target can see the missile coming and dodge it.

Our early ancestors invented lethal missiles to dominate their small world, then added a series of highly sophisticated innovations that enabled them to exploit the environment to levels no other large creature had ever done before.

Yet progress was not restricted to the development of a single item or stratagem in our ancient but very competitive world. In small tribes with a variety of tasks to perform there was a continuous incentive to maximise hunting efficiency.

That incentive was to out breed and out compete the neighbours.

Now that man was the apex predator, his need was for ever more territory to exploit.

In lands in which every family group and tribe could expand without either predators or hunger culling their growing populations, smaller groups had to adapt to live on the margins or face destruction at the hands of their fellow men.

At the time missiles made their mark on prehistory, other inventions further enhanced human dominance. These include pit traps, weaving and woven baskets, string and rope nets or snares. In turn such items were prerequisites for another batch of inventions which made life very much easier for hunter-gatherer tribes.

Years or generations later, humans would combine string-threads with pliable wood to produce bows and arrows that would facilitate hunting and warfare on an unprecedented scale. The bow and arrow was the (land-based) hunter's first-choice weapon for the majority of human history.

Some researchers suggest that the first bows were in use at least 70,000 years ago (and may be an innovation copied from Neanderthals).

Yet while it is true to say that a bow is no more than a stick and a string, the development of the bow was relentless.

Every aspect of its performance was enhanced and its uses adapted to the demands of specialist hunting and eventually to large professional armies.

The resulting projectiles were designed for very specific purposes (different arrowheads were used for hunting different sized prey, casting flame, bursting through thick armour and for shooting at different ranges) and increasingly powerful bows were used either singly or en masse as the situation demanded.

To obtain extra power, bows were at first made from natural hardwoods carefully selected for their strength and flexibility (such as ash or yew), then the central wooden staves gradually strengthened and lengthened, before ever more exotic woods (or horn) were laminated together with strong glues.

The bow-string also underwent improvements. Various oils protected both string and wood from moisture. Even the users were selectively improved, with archer-castes trained for a decade or more to build the raw muscle to power and pull massive war bows.

Arrows could be simple and quickly made or highly sophisticated.
Developments include horn inlays to strengthen the shaft at the point of release, silken thread to reduce weight, various shaft shapes for distance or maximum penetration, different feather lengths and twists to maximise accuracy or penetration and, as

mentioned previously, a huge range of arrow-head shapes and sizes.

So, going back to our modern 'Crusoe' family, they instinctively know how to throw stones and sticks and, thanks to humanity eliminating all major temperate zone predators (at least in the western world), they can easily defend themselves.

In nature, even the most efficient predators don't achieve a hundred percent success rate.

Would our family know enough to feed themselves an adequately protein-rich diet? Would they be skilled enough to avoid injuries? Could they move up a huge technological step and make a working bow?

*And would they survive long enough to find out?*

Great Invention number 2.

# Weaving & Knots

*Developed around three hundred thousand years ago.*

Only a few naturally occurring fibres are long enough to form into or bind any really large objects, although there are ample uses for even short threads (such as fastening hides together or holding spear heads onto their shafts). Perhaps the first great invention to lengthen such twine was the knot.

But it was almost certainly from short threads that someone noticed entwined or tangled materials would stay in place under even quite powerful strain.

Woven or plated rope was born and a plethora of organic materials quickly found and exploited that would significantly improve almost every aspect of human life.

We don't know if early man was weaving baskets much before the invention of rope, but at some point in the distant past woven baskets (made from grasses, sinew, hide strips, flexible bark, hair, vines or even springy wood) were used to collect fruits and nuts and trap fish and crustaceans.

A loose weave basket or the observation of a spider's web may have inspired the creation of the first net.

A basket with bait underneath would have been used to trap birds and tethered in water to trap fish.

We've already noted how the invention of woven ropes would lead to bow-strings and the huge impact that bows had on hunting, defence and warfare.

Rope also enables the lashing together of animal skins, building materials and fixing up such tools as stone axe heads to handles.

Nets gave an immense boost to the quantity of food that hunters could secure, allowing trapping to take place when the hunter wasn't physically present.
Baskets make the collection and transport of small objects, particularly fruits, nuts, berries and seeds, much more efficient.

There will have been periods and places in our history where the amount of physical effort required for a hunter to feed even his extended family could have been counted in minutes per week, rather than the dozens of hours most workers need to put in today.

Less time required for work means more is available for discovery, exploration for its own sake and the contemplation of the natural world that would lead to science, philosophy and religion.

But extra food for us, as with all animals, leads to rapid population growth and the initial advantages to the individual tribesmen are soon lost. Weaker families or tribes are driven to more marginal locations where even the best inventions are no longer enough.

When you can't trap or catch enough food close to shore you have to look further out. You need a base to operate off-shore or reach that next island.
And once you have rope, things that float can be lashed together…

*Our marooned family would know about rope, but would they know enough to gather the necessary raw materials from their environment?*

*Would they know how to soak or boil materials to make them soft and pliable?*

*Could they fashion the products vital for obtaining more food, or build shelters robust enough to withstand storms?*

*Could they keep warm and dry in the long damp winters that inhibited the spread of our distant ancestors for so many generations?*

Great invention number 3.

## Clothing and shelters.

*Initial discovery unknown but not less than 80,000 years ago*

The lack of body-hair is very much a double edged sword. It gives huge advantages to the water-adapted or long distance running ape but demands far more care in temperature regulation and avoiding desiccation or strong direct sunshine.

Therefore, it is no great surprise that protective clothing actually appeared before early humans, the first nearly-naked apes, could venture out of their tropical homeland.

Today, apes and chimps will shelter under leaves and build temporary nests away from predators.

Tropical climes and thick body-hair meant these creatures had no need of higher quality clothing or warmer homes, but when the naked ape ventured outside the tropics, protection from the elements was essential.

Our progression from nests to purpose built shelters was probably no great cognitive leap.

Over the annual cycle of hot and cold and wet and dry periods, most non-tropical environments demand clothing and shelter for a naked ape to survive.

Anthropologists have ample evidence for cave dwelling, but the types of early shelters made by man will include various temporary (leaf or earthen) shelters that wouldn't be preserved adequately to allow modern discovery and analysis.

Once leaves and grasses are used as roofing material, incremental improvements will lead to a fully waterproof thatch. More durable structures include earth and stone piles, mud daubed and/or wooden structures and hides in the extreme conditions of the arctic (most snow-homes use snow on top of the hides for insulation).

Shelters offer some degree of protection from the elements and some security from predation. Early homes were created with restricted access via narrow or block-able entrances, or made high up on hillsides or in trees.
Our pre-human ancestors very soon learned that the creature they needed most protection

from was others of their own species.

Modern homes are really nothing more than artificial caves.

Early artificial caves allowed greater comfort and safety throughout existing territory but didn't offer much of an extension to that geographic area.

Regardless of what shelters are available as a base from which hunter-gatherers can explore their environments, they also needed clothing to move around comfortably outside.

Early clothes in the form of scrapped hides were a feature of Neanderthal families from over two hundred thousand years ago, and while these hominids were probably well ahead of us in the development of clothing (evidenced by the fact that they left the tropics and colonised extremely cold areas of Europe long before modern humans), we either copied them or developed independently similar uses for natural materials.

Woven textiles appear around eighty to one hundred thousand years ago (evidenced by the diversification of the head louse into a body-louse around seventy five thousand years ago).

It is likely that the date for the first modern humans spreading out from a tropical homeland at that time is not just a coincidence.
Clothing probably started with the use of animal skins and furs.

When there was abundant meat, tough hides would have been discarded and left lying around.

Although such neglected hides usually dry out and form hard sheets, a few would have been contaminated with the materials that were found to soften leather. These rather unpleasant materials are animal (including human) dung and brains. Urine helps with the removal of hair from the hide.

Perhaps the first cured and therefore supple hides started life as discarded toilet paper…

Whatever happened, some hides were found to have softened.

Experimentation over years or generations eventually uncovered processes to make the hides even softer, more durable and flexible.

Natural hides are a fantastic resource with uses in clothing and footwear, water-skins, belts, glues, boats and sails and bags. Thinly cut and woven hide makes for strong ropes and thread. A few hides are naturally waterproof, with the addition of vegetable oils easily making most others water resistant.

At some point humans went from softening leather to hardening it to make body-armour (leather body armour is still in use in advanced societies today as a protection for motor cyclists.)

The first use for skins was probably as bedding with the hair left on. The first clothing was probably nothing more elaborate than a large hide with an opening cut into it for the head to go through, making a rudimentary but still very effective cloak or mantel.

Yet the drawback with hides is that few are big enough to cover a full grown man adequately, so it was logical that a means of fastening hides together was quickly adopted.

This was probably done by tying the hides with a simple rope or yarn, possibly using a sharp stone or bone to punch holes in the material.

Another naturally occurring resource found with butchered animal skins is sinew and gut materials, both of which make excellent threads. This would have been the precursor to sewing, a process from which spinning and weaving would soon evolve. As clothing became more tailored and sophisticated smaller holes would be desirable, giving rise to the need for ever smaller needles.

Early naturally-occurring needles include cactus spines, hog bristles, porcupine quills, then early tailors progressed to shaped and sharpened bone, horn or antler and eventually metal.

Yet the spread of humanity throughout the temperate zones required more skills than just weaving and the construction of shelters.

Bands of hunters, perhaps driven by stronger tribes to more marginal land, would have followed seasonal food sources into less hospitable climates. Human territory would be temporarily expanded, but still limited to core habitats that offered a year round food supply and easily accessible water.

We were still restricted to the tropics and the beach by our diet. What was needed was a reliable means of food preservation, especially one that would work well within temperate climates year round, whenever there was a short-term glut of food followed by a shortage.

Perhaps surprisingly, this was readily available to hunters from a time before there even were modern humans.

Humans have made clothing for over seventy thousand years, using grasses, furs and hides before woven material. Staying warm and dry in temperate climates demands clothing, shelter and adequate calories.

*What are the odds against our modern families mastering the required skills in time to survive that first winter?*

Great invention number 4.

# Floats & Boats

*Invented about a million years ago, improved enough for ocean-going vessels by around 80,000 bc.*

The first floating vessels were undoubtedly single log rafts. The early hunters and fishermen would see logs and trees swept along by floodwaters after heavy rains. They will have noticed that some woods float better than others, then when they had access to rope or vines they could easily lash a few logs together to make a fairly stable floating platform.

From a practical point of view, the steps from inshore or riverside fishing to those few extra metres offshore that a raft allows are small, but the cognitive leap, and the huge benefits accruing, were disproportionately large.

Fishermen and crustacean gatherers could access previously inaccessible foods and, by repeatedly returning to the raft to temporarily store their catches, they would become much more productive.

The transition from rafts to boats would have been steady and rapid.

The highly manoeuvrable dug-out canoe appeared around 100,000 years ago, with ever-increasing size that would see canoes capable of carrying dozens of men. Use at sea led to the incredibly stable outrigger design.

Such small boats opened up a host of offshore islands and bird colonies for exploitation, as well as enabling families to homestead islands and improve security. The transporting of goods and people opened up massive opportunities for long distance trade.

By 80,000 years ago they could transport humans across open oceans, giving access to geographically isolated islands and even the continent of Australasia.

Yet, due to the organic nature of boat building materials, the oldest ocean-going vessel known is the Dover Boat, dating to only about 1550 BC. There are two striking features to this find. The first is that it would have been, when complete, a massive seventy feet long, while the second is the technical skill and quality of the build. The boat is made of oak planks lashed together with yew bindings.

Such a boat requires a large number of workers to build and crew, while the build method demands knowledge of carpentry techniques, tools to cut trees, split out, shape and smooth planks, bonding techniques, and knowledge of the best woods to use. Despite around ten metres of the boat surviving, how the boat was propelled is unknown. While oars are most likely, it is also possible that fabric, hide or even wooden sails were in use by then.

The great distances crossed prehistorically may have been in more outrigger-style canoes. These are much lighter in construction than the Dover-boat style but, as mentioned above, reliably stable in a turbulent ocean.

*For a modern family on an island, fishing could be the one activity that would provide a year round food source. Assuming that they mastered lines and nets, a boat would enlarge the area they could access, but would they fashion stone axes to cut down and shape trees, the ropes to lash rafts together and the shaped paddles or sails to control their vessels?*

Would the end product be robust enough to withstand the buffeting ever present on the world's oceans?

Great invention number 5.

# Fire

*Controlled 1,900,000 to 400,000 years ago.*

Fire was a discovery rather than an invention, but designing the tools needed to create fire on demand was an act of true genius.

The early tools involved generating heat with friction, rubbing two sticks together or spinning one against another (later with a bow-like tool that greatly increased the speed at which a stick could be spun).

Once mastered, fire offers many advantages to its users. It directly provides heat, light and smoke.

And indirectly, much improved safety.

It is intrinsically feared by most large predators. Only when man had mastered fire and tamed the wild wolf to stand guard could he, at last, sleep safe and sound at night.
Recent studies of the effects of predators on prey species have turned up some very surprising indirect effects.

The reintroduction of wolves to Yellowstone National Park in the USA found that just their presence changed the stress levels of their prey. Constant vigilance and changed foraging patterns reduced their average weight and led to skittish, nervous behaviour.

Amazingly, just the fear of predation had a bigger effect on herd numbers than the killing itself.

The added security from staying close to a bright fire whenever there were dangerous predators around would have been apparent almost instantly.

By mastering fire, developing limited-access shelters, domesticating wolves and family or tribal teamwork, humans removed themselves almost totally from our position as a prey animal.

The result of man being one of the few animals that can take time out to relax in safety is undoubtedly a major factor in our success.

Fire would facilitate land clearing (through slash and burn farming methods), signalling and eventually propulsion.

Some of the unburnt particulates from an organic fire produce soot (and lamp-black) which can be used for inks and dyes. Smoke can also be used for pest control and fumigation, and even now over 300 types of plants are burnt and their smoke or ashes used to treat disease.

Another spin off from the domestication of fire is soap. In its raw form, soap is just a mixture of ash and fat. Greek mythology has soap as a gift from the gods.

Soap is a powerful antibiotic (as described by the writer of Once a Warrior King, when talking about its introduction to Vietnamese peasants in the 1960's).

Modern hunter-gatherer societies have shown that man uses fire as a weapon, to deprive their enemies of access to food sources, and directly to cause injury and death.

Flame has been used aggressively since the dawn of man.

Evidence for cooked food goes back 1.9 million years (again pre-dating fully modern humans) although it may have taken a million years for humans and proto-humans to fully realise its potential in this regard.

Early man undoubtedly realised that different types of fuel burn more or less efficiently, at different rates and at different temperatures. The smoke given off by different fuels also smelt differently and even had different properties, not least changing the flavour and texture of food.

Heat gives comfort in cold times and cooking affects the taste, durability and nutrient value available from food.

Cooking vegetables generally makes more calories accessible to the human digestive system than raw ones, vitally important whenever there are food shortages.

It is most likely that man's first use of fire in a controlled form was for heating.

*Fire is what allowed season incursions into temperate zones to become permanent.*

Any group of humans sitting around a fire will explore and experiment with the flames and the materials to hand.
It was therefore inevitable that fire's effects on food (changing taste and texture) and the use of sticks as torches were adapted quickly.

Dipping a stick into oil, resin or fat produced a primitive torch or candle, enabling the exploitation of deeper caves as shelters and giving humans some control over interior lighting. Yet, where fire may have made the greatest long term impact upon human population expansion was in facilitating food preservation (important wherever food is abundant seasonally, but scarce at other times).

Creating and sustaining a fire is not as easy as it sounds. Wood needs to be gathered and dried or it produces dangerous levels of smoke. Kindling needs to be bone dry and the right materials found as firelighters (dry sticks, moss etc.).

Easy enough when the sun shines, but how about on that damp and frosty winter's day?

There are a few societies today that make little or no use of fire or even clothing despite living in cold climates (in the bitterly cold southern extreme of South America, for example). These people tend to have evolved shorter, stouter torsos. It is likely that their ancestors gave up such luxuries due to local scarcities after living in such areas long enough for their bodies to adapt.

*Could our modern humans harvest enough flammable material to see them safely through a hard winter?*

Great invention number 6.

# Food Preservation.

*Dates for the development of the key stages in basic food preservation are unknown, but not later than 100,000 years ago.*

Traditionally (for the first million years or so of human history) there was only one effective method of food preservation. This was the air-drying of food.

High moisture content of organic material enables micro-organisms (bacteria and fungi) to grow and thereby spoil the stored items.

Therefore, all that had to be done to enable increased durability of food was to dry it out thoroughly and then keep it dry.

Both smoking and salting are alternative means of solving the moisture problem. Smoking, of itself, does little to preserve anything other than the upper surface of a material. It's the extra heat around the smoke source and moisture-reducing warm air which dries the food and reduces spoilage.

In dry season tropical and desert climes, air-drying is fairly simple.
It was only with the ability to control fire and build shelters that this type of seasonal food storage became possible in damp, temperate zones.

Salt or brine storage is another now almost universal technique that would have been available to humans near naturally occurring salt deposits. Perhaps an early human noticed that dead creatures in a salt-rich environment decayed very slowly.

There are of course a small number of foods that will last for considerable periods with no artificial preservation. Some nuts, unfertilized eggs and honey for example.

There are several species of animals (such as fish, snails, turtle and fowl) that can be kept alive (but restrained) for days or weeks until needed.

Rabbits were deliberately left to infest some islands as a food source for passing ships as early as the seventh century (possibly before they were commercially domesticated).

In any environment where food is not naturally available (or sufficiently abundant) to maintain a viable population of humans on a year round basis, food preservation is absolutely essential.

With food preservation, marginal, seasonal environments can be turned into permanent homes and with the long term habitation of such locations early humans would be able to adapt more fully to the specific locality. They would have the time to master the local 'off-season' food resources not initially apparent.

Once a basis in food preservation is mastered, the population can then develop both additional foods that can be preserved and additional preservation techniques (such as pickling using acids, fermentation to produce alcohols and cooking processes). The downside of food preservation is the need for the storage environment to be very dry and protect the preserve from direct spoilage and theft by insects, rodents, scavengers and competing humans.
Having managed to obtain a food surplus and protect it from spoilage, the next thing needed was a means of transport….

*Preserving food is one of the biggest challenges that our modern family would face. The resources that they need are there (salty seawater most easily obtained), but could they adapt and manage their assets long enough to master seasonal foodstuffs on their new island home?*

*Could you?*

Great invention number 7.

# Trade

*Started around 2.6 million years ago.*

Trade has its roots firmly established in the time of the dawn of man. It effectively gives a group access to resources (and later, skills) from outside the unit's home territory, while the traders themselves facilitate the spread and diffusion of innovations.

Without trade a society cannot really develop, although it may expand by wars of conquest.

As soon as food collection and storage is done efficiently enough to allow people leisure time, individuals can begin to specialise and produce the goods other people want.

We know that flint arrowheads and stone axes were traded in the early stone age. Skilfully made flint and obsidian knapped stone tools appear suddenly in modern day Ethiopia, in the sediment of the Awash River, dating to around 2.6 million years ago.
These early tools are likely to have been produced and used by pre-human Australopithecines or Homo Habilis (no modern Homo sapiens were around then…).

It then takes about a million years or more of archaeological history before we find strong evidence for other stone-age items being traded.

This has led many archaeologists to suggest that early man wasn't creative.

*They could not be more wrong.*

It is likely that the softer, more organic and perishable goods then around were traded in even earlier periods and over equally significant distances.

Unfortunately, but not surprisingly, all evidence for such organic materials has perished for anything more than 20,000 years old.

What evidence we find for organic material movements around that period is rendered problematic by our lack of detailed knowledge of the flora and fauna distributions over that time.

Strong evidence of organised long distance trade exists by 20,000 years ago (when we have obsidian [volcanic glass] blades turning up large distances from the point of manufacture).

Obsidian is extremely useful historically as it occurs naturally in relatively few sites and can be matched to its source by analysis of its chemistry.

In addition to arrowheads and axes, some of the earliest organic materials known to have been traded are the simple gourd (or squash fruit), seashells and beads.

Gourds were most commonly used for carrying water, storing supplies, as musical instruments, and as rattles.

For hunting and fighting, gourds were used as trumpets. They were also made into cooking and eating utensils. With so many uses, it is not surprising that items like gourds were transported outside their original natural areas of distribution.

Gourds may have been the first crop deliberately domesticated.

The role of the gourd in expanding human exploration will never be known, but it is likely to have been huge.

Once humans had managed to preserve food and water they needed to make it accessible and transportable. There were probably very few options available before the invention of ceramics that were waterproof (vital for transporting most foods over any lengthy period or distance) and reasonably comfortable to carry.

What alternatives existed, such as bags made from hide, needed special treatment to make sure that they didn't dry out and split or soak and rot.

Some woods would be suitable but take hours to work into usable shapes with primitive horn, bone or stone scrappers. The gourd was therefore an ideal transportation device as, once dried properly and cleaned out, it was the right shape for carrying various victuals and could even be easily corked to prevent spillage.

With a rope or hide strap it becomes comfortable to transport and, with care such as oil and resin treatment, can be rendered adequately robust.

The original stone-age proto-homo-sapiens was tied to a known water source, which would have to be visited every single day. That gives him a maximum effective range of no more than six to ten miles from his base, or forced him to follow known routes where water was guaranteed.

By adding a transportable supply of water, he can reduce his refill period from daily to about once in three or four days. Even without finding additional water on his travels he has still increased his effective range from a mere 6 miles to over twenty, vastly extending the territory that can be utilised.
This gives the naked ape two massive advantages in his struggle for dominance over planet earth, he can gather far more food from any given base and, therefore, live in much larger communities. In our conflict with other apex predators and competing humans, "God is on the side of the big battalions," as Napoleon observed.

But our marooned family have no-one on their island to trade with. *They need to make everything themselves or build a boat…*

Great invention number 8.

# **Animal Domestication**

*First accepted date is c 30,000 BC, likely to be pushed back to over 70,000 years ago by new discoveries.*

Animals have been adapted by humans to provide a huge range of benefits.
For example:-

| | |
|---|---|
| Hunting aids | *Dogs, Ferrets, Asian Elephants, Hawks, Cormorants* |
| Herding aids | *Dogs* |
| Sentry Duty | *Dogs, Geese* |
| Beasts of burden | *Dogs, Horses, Donkeys, Lama, Alpaca, Cattle, Camels, Elephants* |
| Meat | *Dogs* (a species specifically bread for consumption is the Mexican Hairless Dog), *Sheep, Goats, Cattle, Pigs, Horses, Fowl, Yak, Deer, Fish, Mice, Hamsters and Rabbits, Snails* |
| Milk & Blood | *Sheep, Cattle, Goats, Horses, Reindeer, Alpaca* |
| Wool & Hair | *Alpaca, Sheep, Yak, Goats, Mink, Horses* |
| Honey | *Bee* |
| Pest Control | *Cats, Dogs* |
| Sport (hunting) | *Deer, Fowl, Fish, Boar, Hare* |
| Medical Research | *Various small mammals, Insects, Dogs, Pigs and Primates* |
| Medicine | *Leach, House fly (for its maggots to clean wounds)* |

Plus a huge variety of animals used as companions (pets) and for their hides.

Of the tens of thousands of species of wildlife on earth, there have been surprisingly few successful domestications. All the animals so far domesticated, (with one or two exceptional omnivores, insects and some fish), are herbivores that live in herds, flocks or packs (and thus have a hierarchical pecking order).

As can be seen from the table below, the first domesticated animal was probably the Dog from the Grey Wolf (and/or the Coyote). What is not so obvious is how poor our information about such domestication is.

Recent finds in Belgium & Siberia could push the date of wolf domestication back to over 31,000 years ago, with the first cooperation between men and wolves obviously being many dog-generations before that. Even the 31,000 year age is a potentially vast understatement, for this only refers to wolves transformed by selective breeding into a non-wolf-like dog, identifiable from skeletal remains.

It is likely that the transformation of wolf to dog was extremely slow at first, as humans probably lived in fairly isolated tribes with small numbers of wolf 'helpers' and didn't consciously understand how to breed their wolves selectively for specific adaptations.

What slowed the domestication of dogs was the limit of territorial overlap between the two species before about 80,000 years ago.

Wolves were a forest-dwelling temperate-zone animal, humans a tropical, coastal animal.

Once domesticated, wolves were massively exploited to meet almost the whole range of human needs and would have decisively improved the competitive edge of the groups who first used them.

Looking down the list below, you'll see that most of the animals domesticated are not tropical species, and, where they are found in the tropics, didn't have a major impact upon humanity or human history (with regards to the quantity of additional humans that can be supported).

The impact of the domestication of large (temperate-zone) mammals (cattle, sheep, goats and horses in particular), upon humanity in recent prehistory (ie in the last 10,000 years), has been huge.

The initial consequences, as with the opening up of any new food resource, was to increase local population growth.

Vastly significant to later human history was the introduction to certain humans of animal-born infections (see note at the end of this chapter).

These infections would, much later, see the Americas conquered by Europeans using a devastating germ-warfare system they had no direct control over. (This refers, of course, to the millions of indigenous American Indians killed by the Typhus and Smallpox epidemics brought from the old world to the new in the 16th century.)

What was significant here was where the domestications originated, as this gave the local population resistance to the animal-born infection, not available to populations that had no contact with those breeds.

The impact of beasts of burden upon human development is another of those things which cannot be overstated.

Transport animals (specifically horses and camels), enabled the range of the individual human adventurer to increase from a maximum of about 20 miles per day to over 200, but in practical terms the increase was far more than these 'how far can a man go' statistics might suggest.

A horse can carry a passenger *and* the food and water he needs, while generally not needing to transport food for itself. Realistically, we are looking at the change from a burdened man's comfortable range of about 6 to 10 miles, with a rider's range of thirty to fifty miles and the latter arriving relatively fresh.

Without beasts of burden the development of humanity would have proceeded at a significantly slower pace. Mounted men could exploit vastly increased territories, fight and trade over much greater distances and communicate much more easily throughout their growing empires.
They could also mix their genes with people from much further a field…

*Perhaps, to give our modern family a fighting chance, we should let them take their pet dog…*

| Species | Estimated date of domestication | Location |
|---|---|---|
| Dog (Canis lupus familiaris) | 33000 BC | Northern Europe East Asia Africa |
| Sheep (Ovis orientalis aries) | 11-9000BC | Southwest Asia |
| Pig (Sus scrofa domestica) | 9000 BC | Near East China Germany |
| Goat (Capra aegagrus)hircus) | 8000 BC | Iran |
| Cow (Bos primigenius Taurus) | 8000 BC | India Middle East Sub-Saharan Africa |
| Zebu (Bos primigenius indicus) | 8000 BC | India |
| Cat (Felis catus) | 7500 BC | Cyprus and Near East |
| Chicken (Gallus gallus domesticus) | 6000 BC | India Southeast Asia |
| Guinea pig | 5000 BC | Peru |
| Donkey (Equus africanus asinus) | 5000 BC | Egypt |
| Honey bee (Apis mellifera) | 4000 BC | Multiple locations |
| Domestic duck (Anas platyrhynchos domesticus) | 4000 BC | China |
| Water buffalo (Bubalus bubalis) | 4000 BC | India China |

| | | |
|---|---|---|
| **Horse** (Equus ferus caballus) | 4000 BC | Eurasian Steppes |
| **Dromedary** (Camelus dromedarius) | 4000 BC | Arabia |
| **Llama** (Lama glama) | 3500 BC | Peru |
| **Silkworm** (Bombyx mori) | 3000 BC | China |
| **Reindeer** (Rangifer tarandus) | 3000 BC | Russia |
| **Rock pigeon** (Columba livia) | 3000 BC | Mediterranean Basin |
| **Goose** (Anser anser domesticus) | 3000 BC | Egypt |
| **Bactrian camel** (Camelus bactrianus) | 2500 BC | Central Asia |
| **Yak** (Bos grunniens) | 2500 BC | Tibet |
| **Asian Elephant** (Elephas maximus) | 2000 BC | Indus Valley civilization |
| **Banteng** (Bos javanicus) | Unknown | Southeast Asia, Java Island |
| **Gayal** (Bos gaurus frontalis) | Unknown | Southeast Asia |
| **Alpaca** (Vicugna pacos) | 1500 BC | Peru |
| **Ferret** (Mustela putorius furo) | 1500 BC | Europe |

| | | |
|---|---|---|
| **Fallow Deer** (Dama dama) | 1000 BC | Mediterranean Basin |
| **Muscovy Duck** (Cairina momelanotus) | Unknown | South America |
| **Guineafowl** (Numida meleagris) | Unknown | Africa |
| **Common carp** (Cyprinus carpio haematopterus) | Unknown | East Asia |
| **Domesticated turkey** (Meleagris gallopavo) | 500 BC | Mexico |
| **Goldfish** (derived from Common carp) | Unknown | China |
| **Indian Peafowl** (Pavo cristatus) | 500 BC | India |
| **Barbary Dove** (Streptopelia risoria) | 500 BC | North Africa |
| **European Rabbit** (Oryctolagus cuniculus) | 400 AD | France |

It's likely that one of the first animals kept for its meat was the snail, although it's proving difficult to establish a date.

*A fabulous book covering this area in more detail is,* Guns Germs and Steel *by Jared Diamond.*

Great invention number 9.

# Agriculture

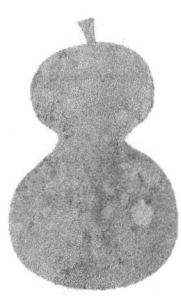

*Established by 10,000 bc*

The emerging consensus is that domesticated plants spread rapidly, but only slowly changed people into farmers.

*Why was this?*

It comes back to the opening scenario of people being given a sudden glut of food. Irregular harvests are not a recipe for long term stability.

Even ten thousand years after the discovery of the first grain farming, up to and slightly beyond the end of the middle ages, farmers regularly experienced serious food shortages in the early northern, temperate summer.
Here, the stable grain crop was either used up or spoilt by a fungus, Ergot. Ergot poisoning was known as St. Anthony's Fire after the burning sensation produced by the constriction of blood vessels and the Medieval monks of the Hospital Order of St. Anthony who cared for victims.

Light use of Ergot-infected grains induces hallucinations, too much kills.
Therefore, it was only practical for mankind to completely give up the hunter-gatherer lifestyle when both good storage systems and a variety of crops were available to facilitate a reasonably healthy year-round diet.

Such a diet is best achieved by mixed farming, where crops come into season at different times (Rhubarb was a popular first crop of the year in dark age Britain). The process of moving from the discovery of the benefits obtained from selectively planting foodstuffs to full reliance on sedentary farming almost certainly took hundreds of generations and thousands of years.

And the invention of ceramics.

Plant domestication and the development of farming was something that opened up a truly phenomenal increase in the quantity of calories available from a given land area.

To put this in perspective, only about 1% of the sunlight falling on a field is converted into plant material, and only about 1% of the resultant plant material is converted by herbivores into meat.

By transferring from a mostly meat and fruit diet to essentially grass-seed eaters, farming can gain almost a hundred times more food per acre than can hunter-gathering populations.

Viewed from any vantage point the environmental impact of farming is the most visible sign of human life on the planet.

Hillsides are terraced, forests cleared, almost every inch of fertile land is parcelled up into fields, wetlands and swamps drained and huge acreages of single crop plantations adorn the landscape.

So the question is, given that taking seeds or cuttings from wild plants and developing land to grow those plants preferentially is not actually rocket science, why did it take humans the best part of ten thousand years to achieve the transition?

There are three major problem areas for understanding the role of farming in human development:

1. Coastal settlement evidence is mostly lost.

2. Identifying and interpreting the evidence for domestication.

3. The fact that farming is both a blessing and a curse.

Finally, there is the mysterious, almost spontaneous, worldwide development of agriculture.

Settlement, arable farming and domesticated plants are difficult to analyse when it comes to understanding human development. Archaeologists need to find firm evidence of settlements to locate, uncover and associate any evidence for farming, but many archaeologists seem to believe farming is a prerequisite for the development of settled communities. Logic strongly suggests people adopt permanent settlement sites wherever there is year-round access to food, which is almost always by rivers and on the coast.

This belief in the need for agriculture before settlement is obviously flawed, as only a consistent food supply is needed for settlements to persist. Once there is such a settlement, seasonal agriculture can supplement the settlement's food supply, paving the way for the development of food storage and preservation.

No creature is known that will abandon home territories and nest sites more often than is absolutely necessary.

The last major ice age removed nearly all evidence for human settlement in temperate latitudes before 18,000 years ago.

That date is also the lowest point in recent history for global sea levels, when they were a staggering 130 meters below those of today.

In geological terms the post-glacial sea level rises have been rapid and huge, so any evidence of early coastal settlements, should they have existed, will have been submerged.

Coastal settlements also suffer from a regular cleansing by storms, tsunamis, erosion, changing river patterns etc., especially in a period when most coastal dwellings would be constructed of impermanent materials such as wood or mud-brick.

Given that the best farmland is always on floodplains and these are almost invariably low lying and coastal, the loss of evidence to these natural causes has been severe.

Underwater archaeology is unfortunately still in its infancy.

A plant is said to be domesticated when its native characteristics are altered such that it cannot grow and reproduce without human intervention, or is no longer found in a natural environment.

Domestication is thought to be the result of the development of a symbiotic relationship between plants and humans, in a process of co-evolution.

Human behaviours evolve to suit and modify the evolution of a particular plant type.

In the simplest form of co-evolution, a human harvests a given plant selectively, based on preferred characteristics such as the largest fruits, and uses the seeds from the largest fruits to plant the next year.

This definition means that identification of the specific point when plant domestication occurs is logically impossible to determine, because a plant is either (already) adapted to suit human needs or is 'wild'.

By this definition we will never know when the first steps were taken down the path that would bring about agriculture.

Evidence of domestication at archaeological sites requires both identification and interpretation. Both areas bring their own complex problems, which essentially come down to putting the finds into context.

Imagine a small settlement (which may be temporary or semi-permanent), in which there are remains of either burnt vegetables or plant matter extracted from middens (both sources can be proof of the use of vegetables by humans, but neither specifically points to agriculture).

How can the archaeologist establish whether the plant finds were gathered from the local territory *without* active farming taking place, or whether they were deliberately (or even accidentally) planted nearby for later harvesting?

The only time you can prove that plant domestication, and therefore active farming, actually occurred is when the plant types found are either not native to the area or have been significantly modified from the wild stock by human intervention, both of which are likely to take years or generations to evolve…

Therefore, the earliest farmers deliberately harvesting unmodified local stocks could not be differentiated from any non-farming community, despite the fact that they may be exploiting their immediate environment more productively than would their neighbours.

Yet early and simple farming techniques bring the actively farming group the advantage of higher food levels and potentially more offspring from any given land area. That is, a massive competitive advantage over local rivals.

While harvesting domesticated crops can significantly increase the food yield from a given area (particularly in the form of starches), there is a range of adverse consequences (risks) arising from any change from hunter-gathering to farming.

These include:

*Famine*, through drought, disease, infestations (rodents and insects in particular) or storm damage. These problems are only seriously felt when communities have become crop-dependent.

*Food storage* system failure.

Diet becomes more limited, with specific *nutrient deficiencies*.

*Disease* & sanitation deficiencies arise due to raised population densities.

*Social stratification* emerges as a political upper class and military elite evolve that can be supported by the agricultural surplus.

Physical *mobility decreases* as populations become tied to the land.

Territoriality and intra-species *fighting increases* as populations actively protect their crops and lands and deny access to outsiders.

*Environmental damage* and degradation arise, through, for example, overgrazing, land clearance (particularly soil erosion, desertification, saltification etc.) and nutrient leaching.

# The worldwide development of agriculture.

*The following table demonstrates how the agriculture of today's main crop plants basically occurred throughout much of the world in the two thousand year period leading up to 7000BC.*

| Plant | Date | Location |
|---|---|---|
| Fig | 9500 BC | Near East |
| Rice | 9000 BC | E. Asia |
| Barley | 8500 BC | Near East |
| Einkorn & Emmer wheat | 8500 BC | Near East |
| Chickpea | 8500 BC | Anatolia |
| Bottle gourd | 8000 BC | Asia |
| Potatoes | 8000 BC | Andes |
| Squash (Cucurbita pepo oviform) | 8000 BC | Central America |
| | 3000 BC | North America |
| Maize | 7000 BC | Central America |
| Millet (Broomcorn) | 6000 BC | East Asia |
| Bread wheat | 6000 BC | Near East |
| Manioc/Cassava | 6000 BC | South America |

| Peanuts | 6000 BC | Andes |
|---|---|---|
| Banana | 5000BC | New Guinea |
| Avocado | 5000BC | Central America |
| Cotton | 5000BC | Southwest Asia |
|  | 4000 BC | Peru |
|  | 3000 BC | Meso-America |
| Chili peppers | 4000 BC | South America |
| Water Melon | 4000 BC | Near East |
| Olives | 4000 BC | Near East |
| Pomegranate | 3500 BC | Iran |
| Hemp | 3500 BC | East Asia |
| Coca | 3000 BC | South America |
| Sunflower | 2600 BC | Central America |
|  | 2000 BC | North America |
| Sweet Potato | 2500 BC | Peru |
| Marsh elder (Iva annua) | 2400 BC | North America |
| Sorghum | 2000 BC | Africa |
| Pearl millet | 1800 BC | Africa |

| Chocolate | 1600 BC | Mexico |
| --- | --- | --- |
| Cheno-podium | 1500 BC | North America |
| Eggplant | 1st century BC | Asia |
| Vanilla | 14th century AD | Central America |
| Macadamia Nut | 1890 | Australia |

This, from the point of view of human innovation, makes almost no sense.

People lived in Africa for tens of thousands of years before 10,000BC and appear to have domesticated almost nothing, yet people spread to the Americas only about 5000 years earlier than our 10,000 BC threshold and were soon churning out grains, tubers and fruits. *How and why did this happen?*

A major global phenomenon that affected the worldwide human settlement pattern was the last ice age, which peaked about 20,000 years ago. It is likely that the decrease in food of all types available to humans at this time saw global populations crashing, perhaps to global levels of no more than a few thousand.
As the planet warmed back up humans spread and proliferated again, with significant population growth in the fertile crescent of the Near East, Eastern Asia and Central and Meso-America.

These were the regions where there were wild varieties of plants and animals that were suitable for domestication. It then took a relatively short time for human hunters to over-exploit the herds of large mammals in these regions and, of necessity, develop the skills and tools needed for sedentary farming.

Farming has therefore been the biggest boost to the quantity of human life, but its effects on the quality of life are much more questionable.

Many hunter-gatherer societies forced to give up 'the old ways', such as native Americans, may resent the transition. After all, as a pleasant way to exist, hunting is generally much more fun than farming….

*The modern family would need to retain and store without spoilage a selection of seeds for spring planting.*

*If they survive their first winter and if hunger doesn't drive them to eat their seed stock, perhaps they could evolve into at least subsistence farmers by the second year.*

Great invention number 10.

# Sanitation

*Development started by about 5,000 bc*

Prior to expansion much beyond the tropics and the extended tribal family unit of no more than a hundred or so individuals, there was no great need for an organised and rapid disposal of human waste.

Humans naturally avoid contact with undried faecal material and decomposing foodstuffs and will normally void their wastes at least 2 metres from where they sleep. Nomadic and semi-nomadic people will also regularly move away from habitation that has been polluted, so that they are, in effect, regularly cleansing their dwellings.

The development of sedentary, temperate farming brought home the deficiencies of primitive hygiene as populations settled and grew. There are four key problems to settled habitation:

1) Village populations became so large that waste began to accumulate close to dwellings. Waste volumes rose to levels at which they started to contaminate drinking water.

2) Living in temperate zones meant that cold and wet nights put pressure upon the natural urge to defecate well away from the sleeping and eating areas.

3) Villages and livestock spreading inland and settling upstream meant that downstream water supplies were contaminated with human and animal waste.

4) Human waste and domestically stored food proved a major spur to rodent population growth and rodent borne infections.

For most of human history, cities have been places where the surplus population from the land *went to die*.

Average life expectancy, where individuals lived close to the waste and infections of their neighbours, was significantly less in urban than rural habitats. The highest risks were to migrants from very small communities moving into a large settlement. Their lack of early exposure to most urban diseases left their immune systems unprepared.

The problem of a highly contaminated water supply was largely masked for many years as drinking fermented liquids (wine, ales and mead) helped disinfect these fluids.

In London, by 1730 AD, over 73% of children died before reaching their fifth birthday, a higher infant mortality rate than is believed for even prehistoric times.

Thus population growth led to unnatural unsanitary practices and to the need for more advanced sanitation. When systems eventually evolved to dispose of human waste hygienically, this lead to further population growth as an increasing number of children survived long enough to breed.

Perhaps the most surprising thing about sanitation is that the key discoveries were truly ancient, but that they then took thousands of years to catch on:-

5,500 years ago settlements in Scara Brae in the Orkneys had indoor, recessed, toilet facilities that drained either under or outside the buildings. It is unknown if these systems were used for solid as well as liquid waste.

At about the same time the Babylonians had cesspits under their houses and squat holes in a dedicated, small toilet room. The Babylonians also used specially made clay pipes to facilitate early plumbing.

In what is now Pakistan (the city of Mohenjo-daro) they had flushable (by washing down with water from clay jars) toilet rooms next to wash-rooms that drained into street sewers and drains.

These drains had many of the features of modern systems including manhole covers and traps for solid waste.

Unfortunately for the people downriver, these sewers still emptied directly into the rivers untreated, and cess-pits were often close to wells.

The Minoan civilization on Crete had the first properly flushable toilet (in the palace of Knossos). About 4,000 years ago it had a sewerage system of clay pipes and terra-cotta that was big enough to walk through.

Copper pipes (with lead & tin soldered joints) were used in Egypt and Palestine for plumbing within a few centuries of the Minoan finds.

By 200 BC the Han emperors of China had toilet facilities including stone seats with armrests and hot and cold piped water.

Yet plumbing skills and basic hygiene faded badly with the fall of the Roman Empire and even as late as 1589 the Royal Court in England issued the following warning:

*"no one, whoever he may be, before, at, or after meals,
Early or late, foul the staircases, corridors; or closets with Urine or other filth."*

*The modern family know more about hygiene than our ancestors and should easily avoid any basic errors. The challenge would come when one or more of the family become ill.*

*Could the stronger members look after their kin without contaminating themselves or their food?*

# Man's route to
# World Conquest

As the examples above illustrate, the inventions that enabled the human success story fall into a need driven sequence:-

Firstly, humans had to dominate their environment physically by becoming the top predator.
Then we had to diversify into ever-increasing geographical areas.
Then exploit these niches productively.
Finally we have had to learn to cope with our own success and vast numbers.

So what of the future?

There are still more lands, seas and other worlds to exploit. We are adapted to living only up to a few hundreds of meters above sea level and not at all below it.
The ever increasing need to learn how to live closely together, hygienically, happily and safely becomes more pressing as world population passes the 7 billion mark, while adequate food creation, storage and preservation is an on-going challenge.

At the time of writing the world's first kilometre high structure is being planned, manned travel to Mars seriously considered and the development of floating cities entering advance conceptualisation.

*This is a story that will never end...*

# Author's Note

There are hundreds of incredible inventions and discoveries not listed above. These include language, numbers and counting, the grindstone or pestle and mortar, the hand axe and hide scrapper, pottery, (horse, camel and elephant) riding, birth control, fermenting crops to make ale, wine & mead, writing, armour, metallurgy, charcoal, road surfacing and walkways and, of course, astronomy.

And there's still no room in this extended list for the wheel.

This is because the wheel only gave a local advantage in productivity (proven by the fact that the Japanese invented and then discarded it, while the Mayan's used wheeled toys but no wheeled machinery as such).

Wheels were not used by desert dwellers until the modern invention of the 4 x 4 and tracked vehicles.

In fact anywhere with soft sand or snow and very steep or densely wooded landscapes, pack animals are a better choice.

The wheel also counts as more of a product development than a great leap forward.

The ancient peoples of Britain and the world's great pyramid builders (be they Central American, Egyptian or South East Asian), probably all used rollers to shift huge stone blocks, while arguably the first true wheel (a grindstone) evolved from the mortars used for smashing nuts or grinding grains - tools we know are not unique to humans.

The inventions listed above are those that enabled people to spread and multiply in ways unattainable before they came along.

By this criteria, the wheel was nothing special.

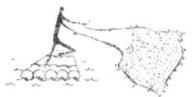

www.ingramcontent.com/pod-product-compliance
Lightning Source LLC
Chambersburg PA
CBHW081752170526
45167CB00009B/4004